STUDY ENGLISH FOR SCIENCE

A.R.BOLITHO

TEACHER'S BOOK

Longman

Longman Group Limited
Longman House, Burnt Mill, Harlow, Essex.

First published 1980
Third impression 1991

ISBN 0 582 74821 6

Printed in Malaysia
by Percetakan Anda Sdn. Bhd.,
Sri Petaling, Kuala Lumpur

CONTENTS *page*

INTRODUCTION

The aims of the course

Students who have worked through to the end of *Learn English for Science* will find in this book a natural continuation. The aim is to bring students at intermediate level up to a point where they can

a) read, understand and take notes on authentic general scientific texts
b) write clear, connected factual English on scientific themes
c) cope with their needs in English without relying on a teacher

By the end of the course, students should be ready to go on to more detailed subject-specific texts in their chosen areas of specialisation. The series as a whole provides a general basic grounding in the language of science and technology.

Study English for Science may, however, be used to advantage on its own by an intermediate student with a sound general educational background as an introduction to the English of Science. Basic numeracy in English is, however, assumed as a starting point.

The book may be used as a course in its own right, or as a supplement to a general English coursebook at intermediate level.

The scientific content of the course.

The book is essentially a language textbook. If study of the texts and exercises stimulates interest in a particular area of science or technology, so much the better. The English teacher, however, will find the material no more daunting than any other language material designed to challenge late intermediate students. Some of the texts are purely factual or descriptive; others deal with problems resulting from rapid developments in technology and industry.

The aim is to arouse interest and reaction in learners of all ages living in a world beset by problems related to rapid advances in science and technology. With a little preparation, there is no reason why even the most unscientific English teacher should not be able to cope with the scientific content. Teachers and students will find it useful to have access to a good general English dictionary (e.g. the *Longman Dictionary of Contemporary English*), a good scientific dictionary (e.g. Flood and West: *An Elementary Scientific and Technical Dictionary* (Longman).

The language used in the course

Early units are more controlled, with some texts specially written and others edited down. Exercises concentrate on important aspects of scientific discourse at text as well as sentence level, on important concepts, on selected structures and on vocabulary building. The texts *are* graded in difficulty, but grading principles have not been applied at the expense of authenticity. By the end of the course, learners should be developing a 'feel' for the distinction between language for active use and passive recognition. There is no need for the vocabulary to be memorized indiscriminately; learners will best remember words which are relevant to their needs. The important language content is in skills such as defining, comparing, contrasting, identifying, classifying and word-building, and in the guided writing and structure practice.

The table on pages 22 and 23 gives a unit by unit summary of the language content of the course.

How each unit is built up

Study English for Science consists of ten teaching units and two revision units. Each unit contains enough basic teaching material for about six lessons of 45 minutes, but follow-up work, project work, etc. may extend this time. Here are some general notes on the individual sections of the units:

The texts

The purpose of the texts is to exemplify different types of writing on scientific, technical and technological subjects. It is hoped that the content of many of the texts will stimulate discussion and reaction which may lead on to project work. Each text is sub-divided into sections for ease of reference and illustrated with diagrams and pictures to aid understanding.

Clearly, there are many different ways of presenting and exploiting texts in a language textbook, and every teacher will have his own preferred approach. Here is one mode of presentation which I have used successfully:

1. (books closed) Teacher introduces the theme of the text by asking questions and finding out what learners know about it, and pre-teaches essential vocabulary and concepts.
2. Teacher writes up two or three key questions or true/false statements (suggestions appear in each unit of the teacher's book) on the board.
3. Learners read text silently, either right through or section by section, looking for answers to questions.
4. Check back on key questions.
5. Learners ask questions about anything which is unclear.
6. Teacher asks detailed questions about the text, section by section.
7. Students summarise text or sections of it, orally and/or in writing.
8. Sections of the text may later be used for dictation practice.

Words

Vocabulary is a problem in scientific texts. The word study sections provide easy access to definitions of some of the key words; where appropriate, examples in context are given. They also provide guidance for the teacher preparing to pre-teach vocabulary. Defining is in itself an important skill for learners to acquire and the definitions in the word study sections offer useful models. In later units, the distinction is made between vocabulary for active use and passive recognition; there is also a problem-solving element in this section in later units. The pronunciation glossary includes all the words included in the word study sections.

There are also exercises on word-building, words in combination and on effective dictionary work. It is worth insisting on the development of this latter skill as many learners remain far too dependent on the teacher for explanations and never learn how to use an English/English dictionary.

Questions on the text

These range from straightforward comprehension questions to true/false exercises and question formation. It is important to give practice in question formation as the teacher is too often the sole initiator of communication in the classroom. Many of the exercises in these sections can be done first in pair or group work.

Sentences and paragraphs
These exercises deal with straightforward grammatical points at sentence level, but also provide practice in discourse features such as connectives, and offer guided writing at paragraph level and beyond. Some of the exercises are on structures, others focus on functions and notions relevant to the English of science. Ideally, these exercises should be taken through in class before they are set for homework.

Graphs
Some exercises under this heading appear in early units. They are concerned with the visual representation of numerical information and with its interpretation.

Intensive reading
There is some extra reading material in later units, always thematically related to the main text. The passages are followed by note-making exercises to encourage efficient reading. They are not intended for close structural analysis or for detailed vocabulary work, but rather to develop the reading skill itself. Most of the texts are authentic.

The revision units
These include exercises on recurring features such as phrasal verbs, prepositions and prefixes, as well as on definitions and verb tenses. They also include comprehension tests. Learners' performance in the revision units should provide the teacher with valuable feedback on their progress towards confident use of English.

The function of this teacher's book is to support your work as a teacher by making suggestions for presentation of material and further activities, and by giving you quick and easy reference to the answers to selected exercises. Only answers to contentious exercises are given. Answers are not given where they can be retrieved *directly* from the text. Nothing about it is prescriptive; every teacher has his own way of presenting material, explaining grammar, etc., and these notes are not intended to infringe this autonomy in any way. They are merely offered to those who wish to consult them.

UNIT 1 The preservation of food

Suggestions for presentation of theme

Bring in plenty of samples of food preserved by the methods dealt with in the text and use them to stimulate students' curiosity about why they don't go bad. Use a map to show immense distances between producers and consumers of many types of food, and a calendar to highlight different times of year at which fresh fruit and vegetables are harvested. Bring in some mouldy bread, sour milk, etc. and ask what makes it go bad.

Extra questions to set before first reading of text

1. Which fungus is useful in certain conditions?
2. In which process is air excluded from food when it is preserved?
3. When food is dried, micro-organisms are killed by the heat of the sun. True or false?
4. Old methods of preservation have been replaced by canning, freezing and radiation. True or false?

Answers to selected exercises

Words

2. (a) length, breadth (b) darkness (c) warmth (d) hardness (e) depth (f) bitterness, sweetness

Questions on the text

4. (a) What have proteins and vitamins in common? (b) Is yeast only harmful? (c) Is canned food boiled before it is sealed? (d) What have salting and smoking in common? (e) What have vinegar and salt in common? (f) How much salt is needed in food to preserve it?

5. (a) False. Only yeast and mould are fungi. (b) False. Fresh food has essential proteins and vitamins which are lost in preservation processes. (c) True. (d) False. Acid works against fermentation. (e) True.

Sentences and paragraphs

6. (a) Smoking and salting are processes used to preserve food. (b) The meat in that lorry is all canned. (c) Proteins are substances necessary for the growth of new cells. (d) Smoking is a way of preserving food familiar to man for centuries. (e) Canning is a way of preserving suitable for most types of food. (f) The food to be preserved is vacuum-sealed in the cans. (g) Malaria is a disease carried by the mosquito. (h) A barometer is an instrument used to measure atmospheric pressure.

7. (a) . . . meat and fish could be kept for long periods if they were salted and smoked. (b) . . . food would not go bad if it was frozen. (c) . . . food would lose most of its moisture if it was dried in the sun. (d) . . . vegetables would lose many of their vital vitamins if they were boiled in water. (e) . . . food would keep longer if it was treated with radiation.

8. (a) Gas liquefies if it is compressed. (b) Water solidifies if it is frozen. (c) Metals expand if they are heated. (d) Metals contract if they are cooled.

(e) Most micro-organisms die if they are boiled. (f) Air rises if it is heated.

9. (a) Petrol is a fuel used in cars. (b) Yeast is a fungus used in baking. (c) Radiation is a process used in food preservation and medicine. (d) Tsetse flies are insects found in some parts of Africa. (e) Glass is a silicon compound used in windows.
10. (a) Dehydration is a process used to dry food. (b) Elephants are animals found in Africa and India. (c) Diamond is a substance used to cut glass. (d) Steel is a substance produced from iron. (e) A catalyst is a substance used to speed up a reaction. (f) Vitamins are vital substances found in foods.

Identification and classification

12. (a) The orange and the lemon belong together because they are both fruits. The potato and the carrot are both vegetables.
    (b) Yeast and mushrooms belong together because they are both fungi. Wine and beer are both alcoholic drinks.
    (c) Tea and coffee belong together because they are both drinks and are both preserved by dehydration. Eggs and onions are both preserved by pickling in vinegar.
    (d) Vinegar and salt belong together because both are preservatives. Bacteria and yeast both turn food bad.
    (e) Liver and yeast extract belong together because they both contain thiamin. Fish and margarine both contain vitamin D.
    (f) Dehydration and smoking belong together because both are methods of preserving food. Vitamins and proteins are vital components of food.

Notes on structures and functions

Exercise 6: These shortened relative sentences occur frequently in scientific writing, and failure to recognise them hinders comprehension.

Exercises 7 and 8: It is assumed that conditional sentences will have been covered before; these exercises are for revision.

Exercise 9: Students at this level must master at least one simple formula for defining.

Suggestions for further activities

1. There is a great deal of scope for project work on food production or on nutrition. Students living in flats in the U.K. cooking for themselves for the first time, could usefully spend time studying the nutritional value of different foods and working out a balanced diet for themselves. The advantage of fresh foods over convenience foods may need stressing! Many producers of preserved foods will happily provide back-up material for projects.

   If the teacher can introduce plenty of tabulated information into project work, similar to that on vitamins in Exercise 11, so much the better. Students of science constantly have to retrieve information from tables, and often have to express it verbally, either orally or in writing.
2. Linked to a project, a visit to a local food factory could increase motivation and stimulate follow-up work.

UNIT 2 Refrigeration and refrigerators

Suggestions for presentation of theme

Use visual aids (e.g. frozen food packets, refrigerator brochures, pictures of refrigerated lorries and cargo ships) to introduce the theme and ask some questions about the advantages of freezing over other methods of preservation. The thematic link with the previous unit is strong, and no special effort will be needed to introduce this material.

Surprisingly few people know how a refrigerator works, and a closer look at its 'working' parts would help to clarify section 4.

Extra questions to set before first reading of text

1. Why is it difficult to freeze most foods successfully?
2. Brine is frozen solid at -40°C. True or false?
3. Which keeps longer at sub-zero temperatures – cheese or eggs?
4. Compression is another word for heat absorption. True or false?

Answers to selected exercises

Words

1. (a) metals – non-metals (b) ferrous compounds – non-ferrous compounds (c) poisonous snakes – non-poisonous snakes (d) toxic substances – non-toxic substances (e) conductors – non-conductors

3. (a) A (b) B (c) A (d) B (e) A (f) A (g) B (h) A (i) B (j) A (k) A (l) B

Questions on the text

5. (a) What is a cell membrane? (b) At what temperature is food fast-frozen? (c) How must frozen food be carried from place to place? (d) Why should food not be re-frozen at home after it has thawed? (e) Which keeps longer, beef or lamb? (f) How long does the compression cycle go on?

6. (a) False. The reverse is true. (b) False. The freezing point of brine is much lower. (c) True. (d) False. Only gases which react in a certain way to pressure can be used.

Sentences and paragraphs

7. (a) Many pre-cooked foods must be heated up before they are eaten. (b) Vegetables should not be boiled for too long. (c) Frozen poultry should be thawed out slowly. (d) Non-acidic foods must be heated up to 120°C during canning or bottling. (e) A refrigerator should be defrosted regularly.

8. (a) A (b) A (c) A (d) B. A freezer is a type of refrigerator we use for long storage. (e) B. Vitamins are substances everybody needs. (f) A (g) B. Farm birds we eat are known as poultry. (h) A

Note that 'which' or 'who' as a subject pronoun cannot be omitted; it is often omitted as an object pronoun.

9. (a) Ships which carry lamb and butter . . . (b) Bottles which contain . . .

(c) Students who arrived . . . (d) . . . in freezers which bear . . .
(e) Vegetables containing . . . (f) . . . booklets listing . . . (g) The thin wall surrounding . . . (h) . . . diseases resulting from . . .

10. (a) The weather has improved gradually over the last week. (b) The temperature rose sharply yesterday. (c) Most metals conduct electricity well. (d) A heart should beat regularly. (e) The two compounds reacted violently. (f) Hydro-electric plants use the power of water fully.

Note that adjectives ending in -l double the -l in the adverb form. Note also that the original sentences, with nouns and adjectives, sound more 'formal' and are more appropriate to scientific writing.

14. Because animal proteins are in fact much more nutritious than vegetable proteins, though vegetable foods shown contain a higher percentage of protein.

Note that this sort of graph is called a **histogram**.

Suggestions for further activities

1. Further work on food projects. Study and description of cyclical processes (e.g. the carbon cycle, the water cycle, etc.). The illustration on page 16 may serve as a starting point.
2. Students with an interest in heat and electricity could find out about different types of thermostat and write descriptions of how they function; the illustration on page 17 may prompt this.

UNIT 3 Spare parts for the human body

Suggestions for presentation of theme

Check on familiarity with parts of the body. Talk about the blood transfusion service in learners' country/countries. Use newspaper reports, etc. on heart transplants and other current medical issues. Biology teachers may have coloured wallcharts to use as visual aids.

Extra questions to set before first reading of text

1. Place these events in chronological order according to the text:
   – The first heart transplant
   – The first fitting of artificial limbs
   – The first blood transfusions
2. If human kidneys fail, does death result due to (a) heart failure (b) poisoning or (c) loss of blood?

Answers to selected exercises

Words

2. (a) blood transfusion (b) kidney machine (c) skin graft (d) blood bank (e) heart surgeon (f) blood pressure

Questions on the text

5. (a) Which organ has never been transplanted? (b) Were artificial limbs and false teeth introduced recently? (c) How can burnt skin be replaced? (d) Who carried out the first heart transplant? (e) Why do many people have to rely on kidney machines?

6. (a) False. Brains cannot be transplanted. (b) True. (c) False. He did tests on dogs first. (d) True. (e) False. Osmosis is a process of absorption of liquid. (f) False. A bottle stopper should be impermeable.

Sentences and paragraphs

7. (a) Most human organs can be replaced. (b) The body may be poisoned if the kidneys are damaged. (c) As far as we know a human brain cannot be transplanted. It may be done one day.

8. (a) It is not possible to transplant limbs but it is possible to replace them. (b) Someone with a bad heart may be given a new one. (c) Blood can be stored in refrigerated blood banks.

9. (a) In South Africa there is a surgeon who has performed successful heart transplants. (b) There are many people who give blood regularly. (c) There are not enough people who give permission for their organs to be used in transplants. (d) There are many people whose lives have been saved by blood transfusions. (e) There are still some organs which cannot be replaced artificially.

10. (a) . . . pump blood around the body. (b) The function of the brain . . . (c) The function of the teeth is to chew food. (d) The function of the lungs . . . (e) The function of a kidney machine . . .

11. (a) A heart-lung machine is designed to replace the heart and lungs. (b) A refrigerator is designed to keep food cool. (c) A slide rule is designed to carry out calculations. (d) A thermostat is designed to regulate temperature. (e) A barometer is designed to measure atmospheric pressure.

12. (a) Kidney machines function on the principle of osmosis, a process by which solids are filtered out of liquids. (b) Tea and coffee are often preserved by dehydration, a process which absorbs moisture from food. (c) A surgeon uses a scalpel, a very sharp knife needed in operations. (d) Meat is rich in proteins, complex chemical substances which give us energy. (e) Operations are carried out by surgeons, doctors who work in hospitals.

13. (a) Vitamins A and D, both (of them) vital to normal growth, are found in fish. (b) Smoking and salting, both (of them) known for centuries, are ways of preserving food. (c) Bread, sugar and potatoes, all (of them) rich in carbohydrates, are part of the diet of most Europeans. (d) Blood donors, many of them young people, save thousands of lives every year.

Suggestions for further activities

Some survey work on successful heart transplant operations – reasons for success or failure of these operations. It may be possible to get a doctor to visit and talk about 'spare part surgery' and the benefits and risks involved.

UNIT 4 Plants

Suggestions for presentation of theme

Bring in, as visual aids, large pictures of flowers, showing pollen. Elicit examples of parasites; discuss effects of some plant diseases (→ famine, food shortages) and of the use of some fertilizers and chemical sprays. The underlying theme is an ecological one, and there ought to be some general talk about food chains and about the balance of nature and the consequences of upsetting it. An aerosol spray, a piece of tree bark, a bottle of disinfectant and a syringe would be useful visual aids for explaining vocabulary.

Extra questions to set before first reading of text

1. What part do insects play in (a) cross-pollination (b) the spreading of plant diseases and (c) the problems of the American robin?
2. Why is it essential to spray crops?

Answers to selected exercises

Words

2. (a) A fungicide is a substance which kills fungi. (b) A herbicide is a substance which kills plants. (c) A pesticide is a substance which kills pests.

3. (a) Meat-eating animals are called carnivores. (b) Mineral-bearing rocks are called ores. (c) Vitamins are health-giving substances. (d) Fleas are blood-eating parasites. (e) Germany is an important steel-producing country.

Questions on the text

5. (a) How is pollen carried from plant to plant? (b) How can plants be killed? (c) What does the type of earthworm mentioned in the text eat? (d) What eats these earthworms? (e) How is crop-spraying carried out?

6. (a) True. (b) False. They also cause plant diseases. (c) True. (d) False. They do not eat the beetles; they are poisoned by eating earthworms.

Sentences and paragraphs

7. (a) Without doctors, many people would die. (b) Without oil, cars would not work. (*or*: Cars would not work without oil.) (c) Without Barnard's work, many people would not be alive today. (d) Without insecticides, pests would spoil crops. (e) Without oxygen, life would not exist. (*or*: Life would not exist without oxygen.)

10. (a) With insecticides, farmers can control pests. (b) With the telescope, Galileo was able to observe the planets more closely. (c) With computers, scientists can work more quickly. (d) With kidney machines, many sick

people can lead normal lives. (e) With the steam engine, man was able to increase industrial production.

Graphs

14

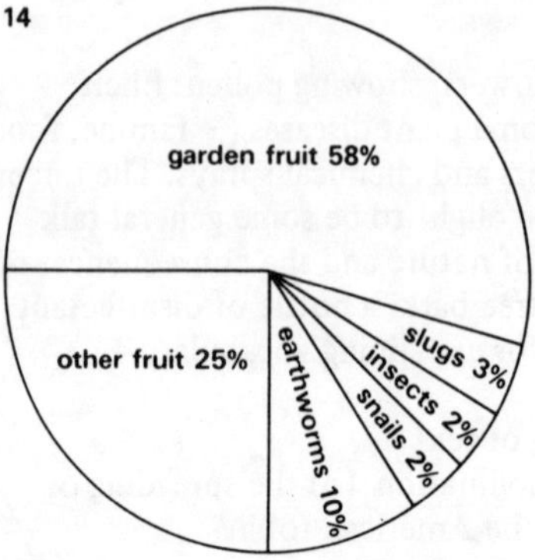

15

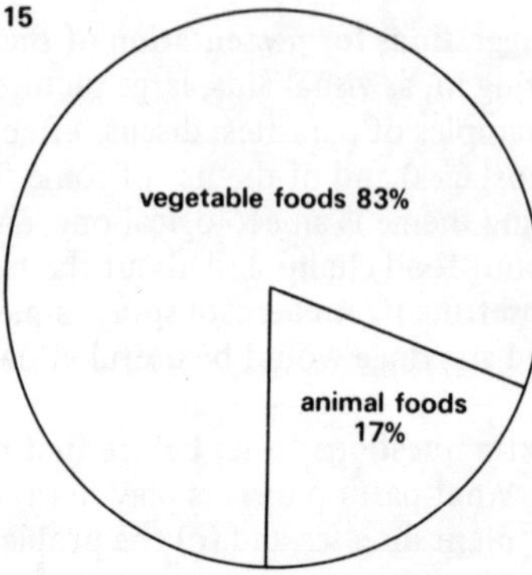

Suggestions for further activities
Students with a particular interest in agriculture and/or botany could follow this interest up in a project on a topic such as 'Plant diseases', 'Crop-spraying', or 'Food chains'. It may be possible to visit a farm and ask questions about crop protection, or to get a farmer or agriculturalist to talk to students.

UNIT 5 Animals in danger

Suggestions for presentation of theme
Talk about animals, particularly those in danger of dying out or which have died out. Bring in maps and pictures of animals. Students from some cultures may be indifferent to animals and may regard the Englishman's concern for animals as puzzlingly sentimental! Accordingly, a good starting point may be to ask about attitudes to animals, and man's effect on animals (industrialisation, pollution, hunting, etc.).

Extra questions to set before first reading of text
1. Which of the following animals is most often written about in newspapers?
   (a) the bison (b) the Komodo dragon (c) the tiger (d) the whale
2. Which animals do these population totals represent? (a) hundreds
   (b) a thousand (c) ten thousand

Answers to selected exercises

Words
1. (a) The treatment of seeds . . . (b) The development of new pesticides . . .

(c) . . . the improvement of food technology. (d) . . . the measurement of the age of fossils.

Questions on the text

3. (a) How many bison were there in North America in 1700? (b) How many Komodo dragons were there in Indonesia in the 1930s? (c) Is the tiger man's enemy or friend? (d) Is the Komodo dragon now found on Rintja, Flores and Komodo? (e) What do the tiger and the Komodo dragon have in common?

4. (a) False. Man threatens animals. (b) True. (c) False. Its numbers were declining. (d) True. (e) True. (f) False. They hunt the animals which the dragons feed on.

Sentences and paragraphs

5. (a) Despite being harmful to fresh food, yeast has some useful properties. (b) Despite being a liquid, mercury is a metal. (c) Despite being banned in many countries, the effects of DDT are still being felt. (*A better version would be:* Despite the fact that DDT is banned in many countries, its effects are still being felt.) (d) The tiger is dying out in the wild despite being protected by law.

6. (a) although (b) although (c) despite

7. (a) . . . efforts to control it. (b) . . . it does not kill spores. (c) . . . advances in medical science. (d) . . . protection. (e) . . . it may die out in the wild. (f) . . . they absorb great quantities of it.

11. (a) There are exactly 168 hours in a week. (b) There are exactly 7 birds in the picture. (c) There are exactly 100 cm in a metre. (d) It weighs exactly 155 grammes.

14. The first map applies to the Indian rhinoceros, the second to the Javan and the third to the Sumatran.

15. The cougars, coyotes and wolves probably fed mainly on weak and sick deer, thus allowing healthy deer to breed. Subsequently, from 1925 onwards, the deer bred indiscriminately, and disease was probably passed on. This must have weakened the species. The destruction of the three species which hunt the deer upset the balance of nature between hunter and hunted.

Suggestions for further activities

1. Students who become interested in this theme could do a project about a selected endangered animal, perhaps one in their home country. The *Red Data Book* is a useful source of information on threatened species.
2. A project on whales could include a listing of all the uses of different parts of a whale together with alternatives which could be used instead.

REVISION UNIT 1 (Units 1–5)

Crossword

Across
1. refrigerators 7. margarine 9. thermostats 13. N.E. 14. sits 16. yard 17. sun 18. cat 19. net 20. lens 22. Cu 23. iron 24. toe 28. react 30. vinegar 32. D.C. 33. leaks 34. Sr

Down
2. fermentation 3. emits 4. arena 5. osmosis 6. semi-conductor 7. maths 8. arm 10. rest 11. oxygen 12. thrust 15. ice 20. long 21. feeds 25. or 26. ova 27. ore 29. A.C. 31. Al

Phrasal verbs
1. (a) died out (b) thawed out (c) stored up (d) runs out (e) switch on (f) comes about

Prefixes
2. (a) transfusion (b) catalyst (c) micro-organisms (d) decimal (e) compressed (f) foresaw
3. (a) transplant (b) substance (c) disinfectant (d) permeable (e) absorbent (f) reproduction

Prepositions
4. (a) on (b) of (c) to (d) in (e) in, of (f) in (g) over

Tenses
5. are; have not existed; have been; have increased; has been; is; has been; have been; has saved

Definitions
All these definitions can be found in Units 1 to 5.

Comprehension Test
9. (a) False. (b) True. (c) False. (d) True.
10. (a) B (b) B (c) A
12. (a) False. (b) True. (c) False. (d) False. (e) True.

UNIT 6 From counting to computers

Suggestions for presentation of theme
The numerals depicted on page 56, the abacus on the same page and the binary and decimal numbers on page 57, can be used to introduce the theme in an interesting way, simply by asking students to cover up the text and say what they know about them. A pocket calculator, a real abacus, a slide rule and some computer cards or tape would be useful visual aids. It may also be useful to enlist the aid of a maths or computer teacher to talk about his subject and particularly to describe how a computer works.

Extra questions to set before first reading of text

1. Which system of writing numerals was more useful, the Roman or the Babylonian? Why?
2. Which came first, mechanical or electronic calculators?
3. Which unit is central to the operations of a computer?

Words

1. (a) digit (b) tribe (c) impulse (d) pulses (e) tax (f) data (g) nerve (h) print out (i) manual

2. (a) a breakdown (b) a build-up (c) splashdown (d) feedback (e) layout

3. (a) intake (b) the outcome (c) an outlet pipe (d) overflow (e) an upthrust

Questions on the text

4. (a) False. They have five fingers, like everyone else. (b) False. Binary maths is taught to help with computer studies later. (c) True. (d) True. (e) False. Information is stored in both the arithmetic unit and the control unit.

Sentences and paragraphs

6. (a) Scientists knew about the harmful effects of DDT in the fifties but it was not until the early seventies that it was banned in Europe.
   (b) Yuri Gagarin was the first man in space (in 1961) but it was not until 1969 that Neil Armstrong landed on the moon.
   (c) The first steam engine was built at the end of the 18th century but it was not until the end of the 19th century that the first motor car was made.

7. Digital computers are really counting machines. They are used for high speed mathematical calculations. With them, numbers can be multiplied, divided, added or subtracted. The smaller ones can be operated manually, like a typewriter. Larger digital computers are programmed with punched cards, punched tape or magnetic tape.

   The information which is put in is read in a unit called the reader. In this unit, the data is converted into a form which the computer can use. These computers are based on very complex electrical circuits.

8. (a) Men used to live in caves but now they live in houses. They used to wear animal skins, but now they wear manufactured clothes. (b) Men used to go hunting for meat but now they buy it from a butcher. (c) Men used to travel on horseback but now they drive cars.

9. (a) People are no longer eaten/boiled alive. (b) Lions are no longer found in Europe. (c) Roman numerals are no longer used in calculations. (d) Scientists no longer use the Fahrenheit scale.

Intensive reading

11. First problem: storage of data for easy access

Solution: data stored on magnetic tapes; selected by computer
Second problem: storage of map outline as numbers
Solution: stored as x and y co-ordinates on magnetic tape, paper tape or punched cards
Method of reading maps: by geameter (outlines recorded as pairs of numbers)
Method of storing data: on magnetic or paper tape, or punched cards
Method of drawing maps: numerical data activates scribing pen which draws outlines
Disadvantage of process: slow and expensive

Mental arithmetic

12. (a) 3½ hours (b) £15 (c) It will cost him less to leave it switched on for a fortnight. (d) 10 minutes (e) He makes £210 per week on milk. He makes £45 per week on beef. His total monthly profit is £1,060.

Suggestions for further activities

1. A visit to a computer terminal would be interesting if it could be arranged, and might result in project work.
2. In co-operation with a maths teacher, you might be able to arrange for some maths classes to be held in English and you could monitor these and work on language mistakes and misunderstandings.

UNIT 7 Sound and noise

Suggestions for presentation of theme

Clearly, a close look at a magnetic tape and at a reel-to-reel tape recorder will help to bring the first section to life. There is certain to be a student in the group who knows how a tape recorder works – let *him* explain it to the rest of the group using a real recorder and simply help him with the English.

One immediately effective way of making the basic distinction between sound and noise is to take in a noisy piece of music, or a recording of, say, a pneumatic drill.

If you have time, collect a tape of all the sounds mentioned in the table on page 66 and see if students can identify them and place them in order of loudness.

Ultrasonic scan tests, which many students may know about, are now regularly carried out on unborn babies inside their mothers' wombs.

Extra questions to set before first reading of text

1. What principle underlies the working of a tape-recorder?
2. Noise affects the ears but no other parts of the body. True or false?
3. Bats produce ultrasonic sound waves for use by scientists. True or false?

Answers to selected exercises

Words

1. (a) amplifier (b) film (c) sonic boom (d) decibel (e) inflammation

(f) vibration (g) magnetostriction (h) quartz (i) foil (j) pasteurization (k) degasification (l) echo-sounding.

2. (a) reverse (b) emerge (c) constrict (d) dilate (e) weld (f) induce

3. (a) nocturnal (b) mental (c) piezoelectric (d) hexagonal

4. *de* – has the meaning of 'take away from'

Questions on the text

5. (a) False. It must be amplified first. (b) False. It is the iron or chrome oxide which is magnetic. (c) True. (d) False. Tapes can be erased and used again. (e) True. (f) True. (g) False. They have many useful applications. (h) False. The frequencies produced by piezoelectric sound transmitters are higher.

8. (a) A piezoelectric sound transmitter consists of a quartz disc cut from a hexagonal quartz crystal and in contact with two metal foils which press on it.
(b) Sound is erased from a tape as it passes the erasing head which is loaded immediately before the recording head and demagnetizes the tape.
(c) People who lose a lot of blood in an operation may be given transfusions from blood banks built up with the help of donors, people who give their blood for those who need it.
(d) Birds eating worms which have eaten leaves sprayed with insecticide containing aldrin, a substance toxic to animals, are likely to die or become ill.

Sentences and paragraphs

10. (a) (i) The basic principle which jet propulsion functions on is that backward thrust produces forward motion.
(ii) The basic principle involved in jet propulsion is the production of forward motion by backward thrust.
(b) The basic principle involved in binary mathematics is that the circuits in an electronic computer are either on or off.
(c) The basic principle involved in the decimal system is that man has ten fingers.
(d) (i) The basic principle which a refrigerator operates on is heat absorption.
(ii) The basic principle involved in the operation of a refrigerator is heat absorption.

11. (a) The preservation of meat, either by salting or by smoking, is a long process. (b) Noise, whether industrial or domestic, is an increasing problem. (c) Ultrasonic waves, whether produced by magnetostriction or by piezoelectricity, have many uses. (d) Many chemical pesticides, both in liquid and in powder form, are poisonous to wild life.

Intensive reading

The sentences in the passage are jumbled. Here is the correct version:

An evaporating basin was taken and half filled with dilute sulphuric acid.

The acid was then heated, but not boiled, over a gentle flame. Copper oxide was added, little by little and with stirring, to the hot acid until reaction stopped. This solution was filtered into another evaporating basin. The filtrate was evaporated gently until a test sample showed that crystals would form on cooling. Then the evaporating basin was set aside and the solution was left to crystallize. Finally, the remaining solution was poured off from the crystals, and these were dried.

$H_2SO_4 + CuO \rightarrow CuSO_4 + H_2O$ – i.e. copper sulphate crystals are formed.

Suggestions for further activities

1. There is plenty of scope for project work here. In Britain, the Noise Abatement Society can supply details of noise levels and their effects. Local surveys of noise problems can be carried out and written up.
2. The work on describing processes can be carried through on any topic relevant to your students' needs – describing experiments, industrial processes, etc.

UNIT 8 Oil

Suggestions for presentation of theme

Using a 'brainstorming' technique, see how many petroleum products your class can name, and how many uses of these products they can list. Oil and petroleum company advertisements from illustrated magazines often show pictures of drilling rigs, tankers, etc., and may provide useful visual aids.

Extra questions to set before first reading of text

1. What have oil, coal and natural gas in common?
2. What traps the oil below the surface of the earth?
3. Where are internal combustion engines found?

Answers to selected exercises

Words

1. (a) blended (b) refined, decay (c) release (d) deprived (e) locate, deposits

Questions on the text

2. (a) True. (b) True. (c) False. Its formation depends on the absence of oxygen. (d) False. Initially, it gushes to the surface under its own pressure. (e) False. Kerosene is more volatile. (f) False. It is a blend of distillates. (g) True. (h) False. Scientists hope to be able to find alternative fuels.

Sentences and paragraphs

4. Students can be asked to try simplifying and shortening this paragraph before referring back to the text.

5. Raising oil from beneath the sea-bed is an immensely difficult and dangerous operation. Though exposed to high winds and heavy seas, drilling engineers have to make test bores to see if it is worth exploring further. Once satisfied that they have found an oilfield, they set up a platform. The quality of the oil piped up to the surface varies, but it all has to be brought ashore, either by pumping it along pipelines or carrying it in tankers. In the North Sea, constantly threatened by the weather, the big oil companies have, on the whole, preferred pipelines. In other locations, favoured by better weather conditions, they often use tankers.

6. (a) The more oil pollution there is, the greater the danger to marine life. (b) The faster oil is used, the higher the price of oil. (c) The higher the temperature in a fractionating column, the fewer substances are left in. (d) The greater the depth of an oilfield, the more difficult it is to raise the oil.

7. An opportunity for individual answers. Check on the correct use of the verb 'take' in answers offered.

8. Another open-ended exercise. Check on the correct use of 'may' in answers offered.

9. (a) The world's oil will have run out by the year 2030. (b) Motor cars will have become museum pieces by the year 2050. (c) Tigers will have died out by the end of the century. (d) Many pollution problems will have been solved by the time our grandchildren are born. (e) Computers will have taken over the world by 1984.

Note that the force of 'by' in these sentences indicates any time before, but no later than, the stated time.

Intensive reading

Fraction	Boiling range	Uses
Gas	–164° – 20°C	Household fuel
Petrol	30° – 200°C	Car fuel
Kerosene	200° – 300°C	Fuel for jet planes
Heavy gas oil	300° – 430°C	Diesel fuel for lorries, etc.
Fuel oil		Redistilled to produce other fractions
Petroleum jelly	350°C	Lubricant; skin care
Paraffin	52° – 57°C	Candles
Pitch and tar	430°C +	Asphalt

Suggestions for further activities

1. The big oil companies (Shell, Esso, B.P.) are usually happy to provide posters and booklets describing the steps in oil production. These can help to stimulate project work, as can a visit to a refinery if there is one near your school or college.
2. If students become interested in the energy crisis, they may enjoy finding out about ways of conserving fuels and in starting a local energy-saving campaign.

UNIT 9 The motor car and pollution

Suggestions for presentation of theme

Talk with the class about the phenomenon of smog and how it differs from fog. Find out how much students know about the pollutants in exhaust fumes; take in a picture or diagram of a car engine and show the cylinders, exhaust pipe, etc. Ask a colleague from the chemistry department to explain the chemical reactions which produce pollutants in a car engine.

Extra questions to set before first reading of text

1. What do the following refer to: (a) 300 days (b) 5% (c) 60% ?
2. What are the major components of smog?

Answers to selected exercises

Words

1. (a) smog (b) an oxidizing agent (c) temperature inversion (d) ultra-violet light (e) ozone (f) haemoglobin (g) additive
2. (Note that other versions may be possible.)

The presence of so many pollutants in the Los Angeles basin means that *comparatively/relatively* little effective farming can be done. Vegetable leaves are soon covered by an *extremely* thick layer of smog deposit and are unfit for consumption. There is a *particularly* clear difference between the condition of the trees on the far side of the hills and those inside the basin. Farming has always been regarded as a *comparatively* healthy outdoor occupation, but tests have shown that the incidence of lung disease among farmers in the area is *particularly/extremely* high, a fact that can only be explained by the dirty air. It is certain that the long-term effects of constant exposure to *highly* toxic emissions such as lead and carbon monoxide can only be harmful.

Questions on the text

3. the geographical location (leading to temperature inversions); the high density of traffic; the effect of sunshine on motor car emissions
4. carbon monoxide; hydrocarbon fragments; nitrogen oxides; lead compounds. Carbon monoxide and lead compounds are directly toxic.
5. (a) False. It is different in that it consists largely of oxidizing agents. (b) True. (c) False. At least, not directly *true*; rich people drive cars which pollute the atmosphere. (d) True. (e) True. (f) False. They have been restricted, not banned.
6. (a) What effect does carbon monoxide have on the blood? (b) What effect can lead have on people? (c) How does sunlight affect exhaust fumes? (d) How does the geographical location of Los Angeles affect the air temperature? (e) What effect does the introduction of more air into the cylinders of a car engine have on the burning of the fuel?
7. These; which; hence; in addition; this; which; thus

Sentences and paragraphs

8. (a) Indian elephants differ from African elephants in that they work for man and have smaller ears.
   (b) A Volkswagen differs from a Mini in that it has its engine at the back.
   (c) A ship differs from a hovercraft in that it floats in the water whereas a hovercraft glides on a cushion of air.
   (d) Kerosene differs from petrol in that it is heavier and is used to fuel jet aircraft, whereas petrol is used as fuel for cars.
   (e) The internal combustion engine differs from the steam engine in that it burns petrol, whereas the steam engine burns solid fuel.

9. (b) high density of traffic. (c) complex chain reactions (d) complex chain reactions (e) oxygen supply to blood cut off (f) prolonged inhalation of carbon monoxide (g) many hydrocarbon fragments left unburned (h) hydrocarbon fragments (i) breathing in of lead particles.

10. These are possible answers. In many cases there are alternative versions. (a) An accumulation of SO_2 in the atmosphere causes normal smog. (b) Smoking causes lung cancer. (c) The effect of exhaust fumes on vegetables at the roadside is to make them unfit to eat. (d) A reduction in the oxygen supply to the blood causes/results in death. (e) Lead additives in petrol have the effect of increasing the octane number.

Intensive reading

12. Hydrocarbons – two types:
   a) Aromatic compounds.
      derivation: coal
      structure: rings of six carbon atoms
      basis: benzene
      two types of reaction: (i) addition of molecules
         e.g. production of cyclohexane
         by addition of six hydrogen atoms
      (ii) substitution of molecules
         e.g. production of nitrobenzene
   b) Aliphatic compounds.
      derivation: petroleum and fatty substances
      structure: chains of carbon atoms
      3 main groups: (i) alkanes
         examples: methane, ethane, propane, petrol, kerosene, oil
         uses : fuels
      (ii) alkenes
         examples: ethylene, propylene
         uses : in margarine and cooking fat
      (iii) alkynes (acetylenes)
         examples: acetylene
         uses : fuel gas used in welding

Suggestions for further activities

There is scope for practical work in motor maintenance, based perhaps on a sourcebook such as *Your car and how it works* by M. Murphy (Longman) – an excellent little reader. Car manuals may also provide interesting detailed material for enthusiasts.

A project could be based on a survey of local atmospheric pollution, perhaps with readings taken at different times of day.

UNIT 10 Tomorrow's world

Suggestions for presentation of theme

'Brainstorming' session on the year 2000, to create a need for accurate language of prediction. These freely elicited predictions can then be compared with the list in section 1 of the reading passage.

Extra questions to set before first reading of text

1. In what ways are scientists concerned with the future?
2. Where did all life begin?
3. Why will insecticides go on being used, in spite of the dangers involved in using them?

Answers to selected exercises

Words

1. (a) prediction (b) speculation (c) nuclear power (d) robots (e) decade (f) diver (g) submersible (h) discard (i) discriminate

Questions on the text

4. (a) True. (b) False. It is at a higher level. (c) False. They do not yet exist. (d) False. There are other means. (e) True.

6. (a) rather than (b) other than (c) rather than (d) rather than (e) other than

7. (a) Lack of money is what prevents progress in space exploration. (b) A world controlled by computers is what many people are afraid of. (c) The safe disposal of nuclear waste is what worries many nuclear physicists. (d) . . . What is known, however, is that the body accumulates large amounts of certain chemicals.

Intensive reading

12. (a) Industry: (i) welding (ii) cutting through hard substances (iii) cutting cloth and piercing leather
    (b) Transport and communications: (i) flight simulation (ii) to replace cables in telecommunications (iii) 'writing' and proof-reading newspapers
    (c) Military uses: (i) 'death ray' (ii) for guiding missiles (iii) range-finding for gunners
    (d) Medicine: (i) to make bloodless incisions (ii) in eye surgery

(iii) to remove skin blemishes [(iv) in cancer surgery]

(e) Measurements: (i) calculating speed of light (ii) measuring distance from moon to earth (iii) measuring site of pollutant particles in air

(f) Pollution control: (i) cleaning up oil slicks on sea (ii) measuring size of pollutant particles

Suggestions for further activities

Students could conduct a survey on fellow-students' speculations about the year 2000 and write a report on their findings. Some may also like to find out and write more about the uses of lasers.

REVISION UNIT 2 (Units 6–10)

Definitions

All these definitions can be found in Units 6 to 10.

Phrasal Verbs

2. (a) breathe in (b) triggered . . . off (c) cutting down (d) pick up (e) helped along (f) cuts off

Prepositions

3. (a) in (b) of (c) from (d) on (e) for (f) in

Comprehension

4. organ; about; caused; convert; vibrations; source; connect; membrane; vibrations; mechanical; transmission; cochlea; causes; move; impulses; that; between; sound; sensitive; from; to; varies

5. (a) False. (b) B (c) B (d) B (e) C (f) C (g) C (h) C (i) D (j) False.

GRAMMAR STRUCTURE SUMMARY CHART

Unit	Branch of Science	Title	Themes	Main remedial structures	Other language work
1	Food technology, biology	The preservation of food	Methods, processes, classification	Contracted relative clauses, 'if' sentences	Definitions; formation of nouns from adjectives; dictionary work
2	Food technology, physics	Refrigeration and refrigerators	Methods, cycles	Passive with 'should' and 'must'; contracted vs non-contractable relative clauses; objectives vs adverbs	Compound nouns; dictionary work; descriptions; interpreting graphs
3	Human biology, medicine	Spare parts for the human body	Functions (of natural organs); systems	Passive with 'may' and 'can'	Connectives (There are . . . who . . .); definitions; giving extra information; step-by-step descriptions of systems; compound nouns; dictionary work
4	Botany, ecology	Plants	Man and the environment; cause and effect (direct and indirect); symptoms, diagnosis and treatment; properties	Negative conditions	Expressing ability; giving examples; interpreting graphs and tables; compound adjectives; dictionary definitions
5	Zoology, ecology	Animals in danger	Man and the environment; cause and effect; numerical relationships; processes;	Negative conditions, comparatives	Connectives (although, despite); comparing; giving estimates; making exact statements; reading for information;

Unit	Branch of Science	Title	Themes	Main remedial structures	Other language work
6	Mathematics, computer science, electronics	From counting to computers	Numerical systems; processes	The passive	Habits; connectives (not . . . until . . .); word building (nouns); reading for information
7	Physics (electricity and magnetism)	Sound and noise	How things work; cause and effect; properties and their applications; processes	Relatives in long sentences	Definitions; expressing principles behind processes; giving alternatives; describing processes; experimental reports
8	Geology, chemistry	Oil	Concepts of time; processes (formation and production)	Past participles in contracted sentences	Expressing related ideas; predictions; dictionary work; reading for information; interpreting graphs
9	Ecology, chemistry	The motor car and pollution	Cause and effect; composition; properties	No remedial structures	Modification of adjectives and adverbs; connectives (in addition, thus, hence, this, these, which); explaining differences; expressing cause and effect; reading for information
10	General (ecology)	Tomorrow's world	Predictions; speculations; hypotheses	Future and future perfect time statements	Making predictions; dictionary work; connectives (rather than, other than); clauses as subjects of sentences; reading for information